SEO Starts Here

Philip Cory

DEDICATION

I would like to dedicate this book to the English teachers in my life. I broke a lot of rules in the book – on purpose.

CONTENTS

ACKNOWLEDGMENTS

I would like to sincerely thank all those who have been a part of my journey over the past ten years. Without the people in my life, I would never have been able to accomplish all the wonderful things that I have. Above all I want to thank my wife, Jessica for her continued support and helping me always stay grounded. I would also like to thank all the people whom I have been able to meet and learn all of these skills from across the United States. This book is dedicated to all those who took the time to teach me everything so I can now share this collection of core basic principles with others.

CHAPTER 1: INTODUCTION TO SEO AND SEARCH ENGINES

To begin I think we should discuss the basic fundamentals of Search Engine Optimization (SEO) and why everyone needs it or wants it. First off, most people or businesses desire to gain more exposure for the purpose of gathering more revenue in one form or another. Simply put, SEO is about money. Therefore SEO is a form of marketing – and nothing else.

For years businesses have relied on other forms of marketing to accomplish these goals. Things such as radio, TV, print ads, billboards, business cards, Yellow Pages, etc. In today's day and age, most of these mediums have succumbed to the power or the Internet and the eyeballs that businesses seek are mostly online.

Think about it for a minute; most of the marketing that you consume is online – not in these old traditional mediums. Sure, we still have a big yellow book occasionally dropped off at our front door. But where does it go? For most people, it's straight to the recycle bin. When you are watching TV, we simply skip through the commercials on our DVR. And the last time someone handed you a business card, where did you put it? Most likely it ended up in the trash shortly after.

That's exactly why we need to find a way to focus online and effectively market to our target audience. Our customers are literally glued to a screen somewhere or somehow during most of their waking hours. Think about your own behavior. How often do you check your email, your social media accounts, or online news? How often do you find yourself streaming video from Netflix or Hulu? That's right! Most people are doing exactly what you are doing. So we will need to explore our own behavior in order to figure out what our target audience is doing.

Now back to the introduction part. I could go on for several pages about the history of the Internet and who started it, where it began and so on. But I will save you the boredom. There are plenty of websites out there that tell that story. To keep it simple – I will merely give you a brief overview of the modern-day Internet and search engines, as we know it.

The first search engine was Excite in 1993. Think about where you were at in the web world at that time. Most of us were not even online yet, at least in our homes. Personally for me, my first online experience was with AOL on a 56k dial up connection. I can just hear the chirps of that modem now!

Along came Yahoo in 1994 as a simple directory of websites. At that time they charged a fee to be included in the directory. (Who knew?) It wasn't until 2002 that they finally implemented a "crawler" similar to what we know these days. Also in 1994 Webcrawler was launched and it included the first full text search feature. Many other search engines cropped up over the next few years through 1997 including AltaVista, Infoseek, Inktomi and AskJeeves.

All hail 1997! Google arrived on the scene from a college-based project out of Stanford University. The project theory was that Internet search could be performed on factors of relevancy and ranking of websites that was based on the links pointing towards them as well as the keywords that were on those sites. As you may already know, Google has become so good at what they do with their ranking systems and algorithms that they now account for over two-thirds of the entire web's searches. That said, I will spend most of my time in this chapter discussing them.

A few others have come along thereafter including Overture, AllTheWeb, and MSN. Many of the aforementioned names played very important roles in the development of search and the web. As with any industry, many of them were competitive with one another and several of them had a lot of cooperation amongst themselves. Ultimately the web as we know it today consists of Google, Yahoo and Bing.

The goal of a search engine is to find and organize information and data that is found online. They do this by using what we refer to as "spiders", "crawlers" and "robots". These whatever-you-want-to-call-them have the job to comb through each page of information and data on the web and put it in to a system of databases. Then the search engine puts each piece of information or data in to large databases and systematically categorizes it. From there an algorithm has been created to rank it versus all the other similar categorized data found online. And upon a query being typed in to the search box of your favorite search engine, results are produced based on that algorithm. Phew! That was mouthful, right? No worries, take your time to digest it and it will make perfect sense. If not, don't sweat it. We will be getting back to some of this in the further reading. This will help you better understand the basic inner-workings of search engines.

So how did SEO come about? Well remember how I talked about the basic premise of it all was to make money? That's exactly what happened. Some folks in the early 90's found that they could use the Internet to actually market online. When this happened, the race was officially started and more people wanted to have their share of this new frontier.

As with any new development, there was a "flying blind" period with no official method of learning. It was basically a lot of trial and error along with many people trying to manipulate or game the algorithm. Those who were working in this online marketing space or early adopting business owners began to compete with one another to try to get better positions in the search engine rankings. Ultimately they were trying to optimize their websites and web pages.

For about the next 10 years or so, Google continued to grow as the leader in the search engine market. This was largely due to a man by the name of Larry Page who developed a formula called PageRank. PageRank was basically a formula that was based on the amount of links that were pointed to a website and which websites were pointed to that site. It was kind of a voting system, if you will. Those who had the most links pointed to their site and/or those who had links from websites with bigger reputation, would win the best search position.

During that time frame, SEO's got away with a lot of nonsense. Many people jumped on the SEO bandwagon and figured out that SEO was actually quite simple and they could game the system very easily. The formula for ranking on a search engine was actually quite miniscule.

1. Launch a website
2. Post up some content that was stuffed with a bunch of related keywords
3. Find out how many links a competitor had
4. Go out and buy more links than that competitor
5. Voila! Your site ranked higher than the competitor

The only problem with that was on the user's end. What people were seeing in the search results became sometimes very daunting, to say the least. You may remember having to comb through a couple of pages of search results and doing a lot of clicking on to websites, then clicking back in your browser. This all because you kept landing on sites that had absolutely nothing to do with what you had searched about.

So Google realized that in order to continue being the leader in the search industry, they needed to ensure that they were giving the user the best possible results when a query was entered. Otherwise, they might risk that user jumping ship and going to another search engine for their queries. This wouldn't be good for them, of course, since they need people to come to their site and search in order to expose those users to the paid ads when queries are entered.

In 2010, Google began to implement quite a few changes that stirred up a lot of confusion for the web world and SEO's. In not-so-many-words, it became a big game of cat and mouse. The Big G started launching and testing several thousand algorithm updates.

The next year they introduced their first series of penalties. This sent most SEO's scrambling and several even left the scene entirely to pursue other careers. Then there were many that who chirped about "SEO is Dead" (many continue with this mantra, by the way.) You will usually find those latter folks selling you on Pay Per Click (PPC) as the only way to succeed online.

The first major penalty update was called Panda. Panda was designed to penalize crappy or "spammy" content. So all that junk out there on the web that was merely taking up space and all those websites that were thinly put together for the purpose of getting you to call a phone number or fill out a form, were instantly hit. Google decided that by penalizing these sites that had content that was thin; not relative or poorly written – they could offer up substantially better results to the user upon their query.

The second major penalty update was called Penguin. Penguin was designed to penalize poor or "spammy" links. And for all of us who hired some SEO, some web guy, or performed DIY SEO on our websites – many of us had a tremendous amount of crummy links pointing to our sites.

What used to work was: A + B = C

A. Keyword Content
B. Links
C. Ranking

So are you seeing the pattern here? Google originally used to index websites based on keywords and links. Now they have come full circle to throw out penalties for those who tried to game the system with these factors.

What works today is completely different, and often times not even related. I'm not saying that you should throw the baby out with the bath water here. What I am saying is that it is critical to understand what used to work and why. And see where we are currently and the direction Google is headed with regards to ranking in the future.

I will discuss a lot about these 2 penalties in later chapters and how you can work towards fixing these issues if you have them with your sites. Along with those two major penalty updates, Google has made several algorithm changes, released a lot of information about best practices and so on. In mid 2013 they introduced a whole new engine and transmission, so-to-speak. This new platform was called Hummingbird.

Hummingbird was designed to basically get in to the user's head. The results it produces are based on the user's intent rather than the actual keywords. It explores the meaning behind a sentence or even carries on a conversation with the person searching. Hummingbird takes in to account things such as location and previous search history. As a result, the users are being rewarded with much better answers to their questions.

Think about how Google has improved for you personally over the recent past. Before you used to have to type in "Movies Regal Cinema Dallas" in hopes of finding the local movie theater. Or did you mean that you wanted to see what movies were playing? Or did you mean something else? Now with Hummingbird, you simply type in "movies" and you get a very valuable results page that shows you movies playing near you, where they are located, etc.

Hummingbird essentially demands that our websites place a bigger emphasis on establishing our brands and being helpful with the content we publish. We need to be more useful and provide a great user experience.

So what has made Google so successful? They have always kept the user first. They somehow have gotten the general public to use their name a verb rather than a noun. When someone asks a question, often times the answer is: "Google it!"

OK, so where are we at now with all this? Depending on where you are at with your websites or your client's websites will determine the best course of action. You may be still stuck in 2009 with a bunch of microsites with a lot of crummy content and a ton of purchased backlinks from crazy directory sites. Or maybe you already got hit with some of these penalties and have begun to make corrections. Or maybe you are just starting out with a fresh site! By understanding the brief history of this all, I can now turn you on to where I believe the current and future state of SEO is and is headed.

In a nutshell, Google is looking for the best of the best. – in every situation and in every search query. Users have learned how to use the internet and how to search. They are no longer typing in "Italian Restaurant Miami Florida", rather it is a simple "Italian Restaurant" from their mobile device and Google does the rest. Users know they can get the absolute best answer on search now. They know that there is some form of location-based search going on behind the scenes. They will also become more and more familiar with conversational-type searching where they can carry on a conversation with the search engine and get exactly what they are looking for.

Many in the SEO world believe that the "10 Blue Links" are going away soon from Google. Those are the ten results you see on each page of the Google results. A lot of this is evidenced by many of the newer search results you are seeing recently. Things such as recipe cards, carousels, entity cards, shopping results and so on. If you are unfamiliar with what I am referring to, try doing a couple of quick searches for "Hotels", "Barack Obama", "Big Hero 6" or "Italian Restaurant". You will begin to see several different styles of results that Google is using versus the standard indexed listing.

SEO is definitely not as easy as it used to be, however it is a lot more simple. There are no more shortcuts in SEO. There are no more magic wands, tips or tricks. Those days are over my friends. If you want to be successful in SEO, you're going to have to do some work.

Several years back the industry adopted three terms that you might be familiar with or have heard. They are White Hat SEO, Black Hat SEO and Grey Hat SEO. These are basically referring back to the old western films where the Good Guys wore White Hats and the Bad Guys wore Black Hats. Naturally the Grey refers to someone who tends to use both techniques. I am only going to talk about White Hat SEO in this book. If you see something somewhere that isn't White Hat, you can probably assume it is Black Hat. And you can also assume that the search engines already know about it and will penalize you for it. Fair enough? Let's move on.

CHAPTER 2: KEYWORDS, GOALS AND COMPETITION

Keywords are basically a word or a set of words in a phrase that you would like to rank for when a user enters a query in a search box. So if you own a flower shop in Chicago, you might want to rank for the keyword "Flower Shop Chicago" or possibly even "Roses Delivery Chicago".

How do you know what keywords you should be ranking for? First things first, hopefully you know your own industry and what people might type in to search boxes to find your goods or services. But it can get quite a bit more involved than that. And actually if you have been nestled in to your industry for quite some time, you may be a little out of the loop when it comes to what your potential customers are actually typing in their search box.

The good news is that there are tremendous amount of tools available for you to use (some paid and some free.) I will mention a few of them here to give you a good starting point. But I will say that there are always new tools and websites cropping up with the latest and greatest. So you will want to play around with a few of them and find which ones suit you best in terms of usability and cost.

Here are some of the ones that I will mention. I recommend

signing up for free trials where possible and giving them a test drive. Once you find which ones you like and which ones you feel comfortable paying for, then move forward with your tool(s) of choice.

Spyfu

www.spyfu.com

Wordtracker

www.wordtracker.com

Wordstream

www.wordstream.com

Submit Express Analyzer

www.submitexpress.com/analyzer

SEOChat

www.tools.seochat.com

You should also try doing a few searches on your own and see what kind of results come up. Also be sure to pay attention to the suggestions that are given by the search engine.

Don't forget to use other search engines for this technique. You can get a lot of great keyword suggestions from sites like YouTube and many more. Unfortunately Google got rid of Keyword Tool a little while back. This used to be our go-to tool. And with Google Analytics showing more and more "not provided" it makes it more difficult to easily grab your current keywords. But don't fret because there are plenty of places to gain a lot of insight and data for your website.

OK, so once you have signed up for a few of these and checked your website on these tools, you should be able to come up with a list of keywords that your site currently rank for. Secondly, you should be able to muster up another list of keywords that you desire to rank for. I would recommend you separate these in to two separate Excel spreadsheets for now.

On your current keywords list, you will want to have one column showing the keyword and the second column showing what the ranking position is. You might want to keep this data labeled properly so you can look back over it in the future and do comparisons and see trends and movements.

Now let's take a look at your list of desired keywords. Hopefully you have made a list of several keywords. Once you have this, I would recommend that you do a search for each one. Look to see who is currently ranking for each keyword and where they rank. Maybe keep track of the first 5 positions on Google for each keyword. Take note of the site that is ranking on your list.

Continue on with each of the next desired keywords and see who is ranking for them. Chances are you might begin to see a pattern of certain competitor's websites that are earning the top spots. This is good information and you should be taking notes on them.

From here you will begin to determine who your biggest competitors are. This can also be done with a few tools out on the market today. Nonetheless, I have always found it best to do a few raw personal searches on my own without the help of all the tools. I like to get the real feel for exactly who is ranking in my industry.

Once you have made a list of several of your main competitors, it's time to take a look at their websites and their keywords. It's a good idea to see what they're up to. This way we can have an idea of what it takes to earn those same spots in the search ranking, and possibly what it is going to take to rank better than them.

Having said all this about keywords and how to find them and use them, I would feel it an injustice if I didn't mention the direction the search engines have been moving recently. I believe that search is more focused on the user's intent rather than specific keywords. And the search engines are getting much smarter in rapid fashion. So I still believe that you should compile your keywords and utilize that list for many reasons to produce content for your site. But I should mention that the idea of Topical Targeting is quite possibly one of the more relative issues. Topical Targeting is about creating pages that have a better user experience because your pages don't just talk about one thing and then the user has to find and click on another page to find the next answer to his question, and so on.

By creating page content that discusses an entire topic, you have the opportunity to engage the user by having them read more and possibly start to trust your site. Therefore they may take a call to action that you have laid out for them. And this equals conversions, or money in the bank.

As a takeaway from this chapter:

Remember that you should not only spend a good amount of time analyzing your own site(s), but also your competition. One of the key principles in SEO is that it's not only about what you are doing, but also about what everyone else around you is doing at the same time.

Obviously you can only control what happens with your own site. But by keeping an eye on what the other sites are doing, you can be sure to understand what factors they are implementing to give the search engine a better decision when ranking.

Compile your keywords and then organize them in to categories. Set yourself up to create content around entire topics, not just one or two keywords. Think about the user at all times. What would you want to see on a page?

3 SITE LAYOUT, DESIGN AND STRUCTURE

As a prelude to this chapter, I want to stress how important this part is. Pay very close attention to the following as it is a very critical part of SEO today and in the future.

I have a saying that I use with all my SEO projects:

"You have to clean up your backyard before you invite everyone over to the BBQ."

That means that if you are getting ready to have a party in your house, and you want people to be impressed (and hopefully return to your home), then you might want to think about sprucing up the place prior to the engagement.

There is absolutely no point in putting up a crummy website that has spammy content, sloppy code and design, and then going out and trying to promote it with backlinks and/or paid ads. You will probably get some visits initially and end up with no desired conversions.

The idea here is to make a plan before you even start. Instead of asking the questions like: "How will Google look at my site?" Ask yourself: "What would my visitors want to see?

You never want to build a site for machines or robots. Rather you should be building your site for real humans. Moreover you should be creating it for the needs and wants of your target audience.

A good way to learn about building the nice proper site is to pay attention to what the big sites are doing (like Amazon, PayPal, eBay, Target, etc.) These companies spend a tremendous amount of money and resources to capture the most traffic and the most conversions. Take advantage of what they are showing you for free!

Look at how their site is laid out and how they utilize their content and how it is written. How are they using images? How are they using little things to help guide the user? All-too-often we go to these big sites on a regular basis and forget to look at what they are doing and why. Take advantage of your own experience! You would be surprised what you will see when you begin to look at the psychology behind some of their little tricks. And you might be surprised how easily you can implement the same things or something very similar to your own site to accomplish the same result.

User Experience

The overall idea with site layout and design is the User Experience (UX). When you put yourself in the shoes of your audience, what expectations do you have? What will motivate you to buy now? What will win your trust with this company? Is this site easy to navigate and understand? Can I get what I want, when I want it, and how I want it?

The cool part about UX is that Google is paying attention to what our audience is doing on our sites. They know what pages they came to, how long they were on that page, what links they clicked, and so on. All of these pieces are becoming more of the ranking factors in their algorithm. Why? Because they assume that if someone has spent a good amount of time on a site and clicks around, then they are most likely getting a good user experience and the information on that site is good. Therefore they tend to want to show that site to more people who search for what that site has on it.

In contrast, when Google sends someone to a site and they don't stay on the site too long or they bounce quickly – the user behavior is gathered there as well and taken in to account on the following rankings. Why would Google want to continually send people to a site that is not giving a good user experience with great information?

The tough part about UX is that it is very difficult to attribute any value to it in the beginning. Therefore presenting it with hopes for a return on investment is not an easy sale. That being said, we need to focus on some basic things in order to begin our quest for the best UX

Layout and Navigation:

You will want to look at how your site is laid out. Does it make sense? Can the average person within your target audience understand the navigation and how to get where he/she wants to go? Keep in mind that you are speaking with your audience, not at them.

Often times we create a layout for our site based on a designer's suggestion, a template, or simply what we think it should look like. Once again I would encourage you to look at what some of the big sites are doing as well as your biggest competitors. They obviously have something that is working and Google is rewarding them. So it is wise to use some of their techniques in your layout.

Be sure to look at how the structure is of your URL's. Are you using a common xyz.com/about-us.html or are you using something a bit more ambiguous like xyz.com/?p=8

You will want to take in to consideration the categories of your content and try to silo them where possible. For example, if your website is about iPhones, then you might be creating content that has to do with the iPhone 3, iPhone 4, iPhone 5, and iPhone 6. Therefore you might want to create categories for each of the different models. So your URL structure might look something like this:

iPhone 3: xyz.com/iphone/3

iPhone 4: xyz.com/iphone/4

iPhone 5: xyz.com/iphone/5

iPhone 6: xyz.com/iphone/6

This would clearly give you a structure to discuss the iPhone in general, and then drill down the content to each model thereafter. And the user would be able to easily navigate to their desired model with these URL's.

And overall the search engine would likely be able to understand what you site is about and how it is laid out. Therefore they might be a bit more likely to rank you higher than your competitor because you provided the user with a better experience.

Clean Code:

I can't tell you how many sites I have been through that simply have sloppy written code. (Some of my own to be quite honest.) I once had a site that had a WYSIWYG editor for the text content and I used to write all my content using Microsoft Word. Then I would simply copy and paste the document in to they WSIWYG editor. Without even knowing, all the macros from the Word Document ended up in my code! It was a total disaster that went unnoticed for a couple of years. When it came to my attention, it took me about 30 days to fully unwind all that nonsense from the code of that 500 page site.

When building your site or having a developer do it for you, it is critical for you to review the style of code that is being used. Many people use popular development tools such as Wordpress, which include plugins. Often times those plugins can create a mash up of unwanted sloppy code within your site.

I would recommend that you try to do a decent review of your site's code prior to launch or prior to any SEO efforts moving forward. This way you can try to eliminate any errors sooner rather than later. It may be more difficult later once you have implemented more changes to your site, or the search engines could downgrade your site's rankings, even worse they could penalize the site.

Basic Fundamentals:

Never forget the basics of SEO. I have seen it a thousand times and I have even made these mistakes myself a time or two. But there are a few things that are an absolute must when launching a new site or a new SEO campaign.

First you should make sure you have created and installed a robots.txt file for your site. If you are unfamiliar with what this file is, you can easily do a quick search for it and there are plenty of free robots.txt generators out there on the web. But basically it is a file that is uploaded to the host server that gives instructions to the search engine's robots (crawlers) as to which files it should and should not visit.

Next you should make sure that you have an htaccess file uploaded to your server. This file creates permissions for the proper access of certain files on the server. Just like the robots.txt file, if you are unfamiliar with how to create one, there are plenty of free resources online to quickly put one together for you.

Lastly I would suggest that you make sure to have a good XML Sitemap file. This file is critical for several reasons, but in simplicity it is a file that tells the search engine about each page on your website and how your content is organized. This file can also contain metadata information about a web page like the importance of the page to the overall site, the last time it was updated and how often it changes.

Most of these will also come in very handy when we get in to Google Webmaster Tools.

Other things to pay attention to when you are working on your site's layout, design and structure are:

Title Tags

Title Tags could be considered one of the most important factors on a website. This is because it is often used in so many places.

First it is generally seen in the top of the user's browser. This may or may not be seen by them, but it does give the page an overall label.

Next, it shows up in the results page of the search engine. Therefore it can sometimes enter the equation of the query for the user. In our "Chicago pizza" example, you are likely to find one of the top results is:

Chicago's Pizza | Order Online 'til 5am! Delivery, Takeout…

There are other places on the web that your Title Tag may appear. Nonetheless, it is a critical element to be sure to optimize for.

H Tags

To understand H Tags, think about an old newspaper. They had a headline (H1), which told you what the article was about. So you knew what to expect before you read the article. Then let's say the article was broken up in to a couple of sections which had little subtitles (H2). These told you what each section was about. And they were pretty handy if you were merely trying to skim the article and read only what you wanted to read to get the gist of it.

Common errors with using H Tags are to use them simply for styling. It may seem like an easy fix, but in reality it can have a devastating effect on your overall SEO.

Many have also made the mistake of using too many of one tag. The user and the search engine only want to see one headline. So be sure to only use one H1 Tag.

Remember you are trying to create each page for the user to best understand your content and ultimately buy from you. Don't confuse them with a lot of H Tags.

Optimally your Title Tag should be right around 55 characters.

Alt Image Tags

These are very critical because a search engine cannot actually "see" your images. (Although facial and entity recognition is evolving fast right now and this writing could be antiquated as soon as it is published.) As smart as they are, they still aren't human.

Interestingly enough, they are tasked with trying to find something for humans. So we have to help the search engines "see" what are images are about. We do this by adding in little descriptions and titles about the image in an effort to be helpful. So while you want to be descriptive, you don't want to write a novel about the image. Conversely you also don't want to rely on the file name of the image.

So if you have a picture of a pizza and the file name is simply "pizza", then you might want to expand on that a little. You may want to write the alt text as something like: 14" Deep Dish Pepperoni Pizza

This would give the search engine a little better clue as to what the image is about. And when someone searches for a "deep dish pepperoni pizza", your image might just come up. And if that happens, you might get some nice, relative traffic to your website. In turn, you might even sell a pizza!

Meta Data

Way back in the beginning, this used to be one of the most overused tricks in the book. People would simply put Brittany Spears or Viagra in to their meta descriptions and meta tags and hold high hopes for getting more traffic. Well the idea of getting "more traffic" is not necessarily a good thing, by the way. We are looking for targeted traffic that is users who are looking for what we have to offer and are hopefully ready to buy what we are selling. Just getting a ton of new traffic to a website is basically pointless.

Meta tags and descriptions are no longer a factor to the search engines, however they can provide enhancement for the user experience. By giving a decent description about what the page is about. Often times (not always) these descriptions are used on the results page of the search engine, right below the Title Tag, as a preview for the user.

You should ideally write these descriptions at around 155 characters. Each page should have it's own unique description written. You might also consider that this is an excellent opportunity to use some salesmanship to attract the search user to click on your link instead of the next one.

Keywords in Content

The main idea here is not necessarily how many times a keyword is used, but rather where the keywords are strategically placed within the content. You certainly don't want to stuff you content with nonsense keywords over and over. Imagine being the person trying to read it. You wouldn't want stick around to read an article that was stuffed.

So Google has gotten wise to this and it can basically read your content and comprehend it. Then they can assign it a "weight" or ranking. So if you write a page about Pepperoni Pizza, you had probably do a pretty good job of offering up something pretty unique, but still using the keywords "Pepperoni Pizza" strategically. This being a very difficult keyword to rank for because you have to compete with all the big and little pizza places along with everyone who is offering recipes for it. So don't think you are going to just put up a page about a certain keyword and it instantly should be recognized and ranked by Google.

Remember that SEO takes time. Just because you think your web page should be on page 1 of Google, doesn't mean that everyone else thinks it should. You should be very careful to consider each piece of content on your pages to ensure that you are giving the potential visitor the best information and the best experience possible. That is the only way to have any hopes for ranking.

Forget stuffing your keywords all together. Start focusing on building great pages that are about topics that include keywords and key phrases that are authoritative and unique. Stand out in the crowd and you have a chance to win. If you do what everyone else is doing, you will likely get the same results that everyone is getting.

The keyword discussion can be a slippery slope because we have to write content for humans, yet we have to attract the search engines to our content at the same time. If you are more focused on the reader, you will likely have a better outcome in the long run.

Appropriate Depth of Content

I believe this is another somewhat slippery slope. Depth and length of content has been batted both directions for years. Some used to say: write small articles and blog posts and post them up. This way the search engine will see a robust site. And others have said: write long daunting articles that contain everything there is to know on said topic.

I believe the appropriate length of an article or blog should be exactly how long it takes to inform, educate or entertain someone in that piece. If you have done a good job in 400 words, then so be it. Publish it. On the other hand, if you need to go more in-depth and write 2000 words to give the user the best understand, more power to you.

I think the major factors that the search engine will take in to account on your length of articles is how long people spend on your page, if they click deeper in to your site, or if that page gets shared and/or linked to.

There is no possible way that anyone could tell you what the optimal length of an article should be. For example, how many words would it take you to write about "how to tie your shoe?" Exactly, so why would anyone publish a 2000 word article about How To Tie Your Shoe? It simply doesn't need to be that long.

When I wrote this book, I never considered how long I wanted it to be. I just sat down with my outline and started to write it. My belief was that it would be exactly how many pages it needed to be in order to say what it needed to say. I think we are almost half way through, so we will see where it ends up!

Links Used On Pages (Both Internal and External)

Internal links are the ones that you create that go from one page on your site to another page on your site. These links should be helpful to the user so they can navigate to a related page or help them understand the hierarchy of your website. They can also be used to "tip off" the search engine about another page that is related to the content on that page. Be sure to use them wisely and always keep the user in mind, not the search engine.

External links are the ones that you create to go from a page on your website to another person's website. These are typically used to be helpful to the user in showing them a source or a more authoritative resource on the topic. You should use them with care, as you don't want to pass on too much juice to another website or risk losing a visitor to another website once you have roped them in to yours.

Some of the most common external links are to social media sites like Facebook and Twitter. Often times many websites want to have their social icons represented on their site and they don't even realize they are potentially losing their visitors to these sites. After all, you work so hard to get people to come to your site, and almost instantly they vanish to go play on Facebook. It's certainly something to consider and I will go in to that in more detail in the Social Media section of this book.

The bottom line is that external links can be necessary. But be sure that you are aware of who you are linking to. You are essentially making a claim that your web page content is related to the site you are linking to. This can either help you or hurt you depending on the context of the link. So when someone asks you to place a link or trade links with you, you might want to think it all the way through first.

Mobile:

Lastly in this chapter let's talk about mobile. Not too long ago we heard many saying that mobile is the future. I've got news for you – the future is now.

Mobile devices are clearly overtaking the web and our lives. It most certainly has taken over search.

What does that mean for your business or website? Quite simply it means that you had better be considering the way you deliver your website and content to your audience.

Regardless of the latest statistics of how many people are using smartphones or tablets, one thing is for sure – it's growing rapidly. Many people are finding that their smartphone is the only necessary device for accessing the web. As a result, we need to be delivering our websites and other web assets in such a way that our audience can consume them. As a nudge to steer you in this direction, Google recently stated they are rewarding those sites that are mobile-friendly. There are so many more reasons why you should be mobile ready. The future is now.

What does it mean to make your site mobile responsive or mobile ready? Basically you have to see about how your website looks and navigates on a phone or tablet. There are several sites available to test your site's output on different devices, but I think the best test is to simply go check it out for yourself. Look at all the devices you can possibly get your hands on. Ask friends and family members to help you by looking at your site on their devices that are different than yours.

The basic rule of thumb here is to make your site easy to use and easy to navigate for the visitor. Be sure to use a clean design that is easy to understand. It is all about usability.

Make sure they can access the same content that is available on the desktop version. Try to ensure that the page load time is optimal for them, since many people are often confined to their cell provider's data speed. I would strongly recommend you look in to using a Content Delivery Network (CDN.) You can find one that is rather affordable and it will definitely pay for itself in the rewards you get back from happy visitors and happy search engines. The CDN will also improve the overall page speed or load time on all versions of your site.

Remember to keep the user in mind at all times when developing the mobile version of your site. Think about where people are likely to tap on their screen. Think about where their eyes are likely to be drawn.

Consider all factors on your mobile site including clickable phone numbers. Make your contact forms short and easy to complete. Consider making them pull data automatically using auto fill. This way the user doesn't have to worry about typing in all their data.

Think about who might be visiting the mobile version of your website. They may need a quick way to get directions to your place of business. Or possibly they would like to place a quick order from their smartphone for one of your products, and they are looking for the easiest checkout process. Regardless of their motivation, help them accomplish what they are there to do and you will win.

CHAPTER 4: CONTENT AND HAVING A GAME PLAN

I have already discussed quite a bit with regards to content, but now I am going to go a little more in-depth. Earlier in the book I mentioned that I would talk more about the Google Panda penalty.

Panda

Be sure to focus in on this chapter and you will find the solution to either repair or avoid this. Content is a major part of the equation and basically it is what makes up the over web as we know it.

Almost everything can be considered content. At it's core, we commonly know content to be all the text we see on websites. But it goes much further to include images, videos, infographics, press releases, eBooks, newsletters, slides, blogs, vlogs, guides, whitepapers, reviews, lists, research or studies, and podcasts. So how do we take in to account each of these types of content and integrate them in to our overall strategy?

The answer generally lies in: who you audience is. We have to consider the following:

1. Who we are targeting?
2. Where are they located?
3. What device are they using?
4. What is their attention span?
5. How do they prefer to consume content?

There is a big difference between the 20 year old male who is looking for hacks to the latest video game and the 55 year old male who is searching for investment strategies. Sure both are male and may have certain things in common, however they are both on very different paths and seeking completely different content.

The earlier example might have the shorter attention span and simply desire to get a quick answer to solve the next level on his video game. He might be more inclined to watch a 30 second YouTube video that has his answer.

The latter example might be more apt to read a longer in-depth research document that he can download in a PDF format.

Understand that two completely different audiences call for two completely different strategies. So you will need to really identify you target audience and consider all factors before embarking on your content creation.

Remember that the idea behind your content needs to be about talking WITH your audience, not AT your audience. You need to engage them in order to gain trust and show your authority on the topic. You cannot simply write a page about something just to have a page about something. That has been done time and time again. As a result, the web is full of duplicated spam! This is exactly why the search engines have a ranking system. They filter out all the pages that have said the same thing without anything unique and those that are simply not helpful. And they have rewarded those pages that are informative, unique and engaging. Furthermore, with the behavioral tracking that has been implemented recently, the search engines can tell exactly how good a page's content is based on how the visitors are responding to it.

So what does that mean for you? It means that you should really take an honest look at the content you are producing and think to yourself: Would I read this? Would I trust this business? Would I click further? Would I fill out their contact form? Would I call this company and try to do business with them? If your answer is no, then you should not post that content.

If your content is basically saying the same thing that another authoritative website is saying, why would Google have any reason to rank you on page 1? They already have that spot filled and they are serving it up to users who are showing good behavior feedback. They have no reason to even consider your page or content.

On the other hand, if you have something new, unique or more informative, educational or entertaining – then you quite possibly deserve the consideration of the search engine. At which point, they will probably consider you and give you a test run. Upon that test run, your page will be under the scrutiny of the user behavior. If your page gets better user behavior (i.e. longer time on page, more engagements), then you stand a chance at winning the better ranking than the previous champion. And if your statistics prove not to be as good as the champ, you can be assured your page will rank lower than theirs. It's really that simple!

You see, content is king. It always has been. Google and the other search engines have never made any confusion of this. They simply want you to create something that adds value to the web. The only way to add value to the web is to inform, educate or entertain. Anything else is merely web junk to them.

Have you ever wondered about a search result that renders 872,000 results? Who would ever look that far? Likely no one I would assume. The only thing that matters is being Number One, on Page One. Many statistics have been rallied over the years about the percentages of people who click on results of the first page versus the latter pages.

Additionally there have been numbers thrown around about how many click on the first result versus the latter results. Regardless of what the numbers are (or change to), I know with certainty that the most clicks go to Number One on Page One. Now more than ever, the pages that are in those top spots are the ones who have the absolute best content.

Each piece of content must have these 7 things:

1. Purpose
    (inform, educate, or entertain)
2. Audience
    (who are they, where are they, what device are they on)
3. Schedule
    (when will this be posted, when will my audience most likely consume this content)
4. Platform
    (your website, your blog, social media, YouTube, SlideShare)
5. Originality
    (has this already been written, am I saying it differently, how is my content unique)
6. Authority
    (why should my audience listen to me, would I trust what I am reading, have I ever heard of this company before)
7. Engagement

(will this content make me feel warm and fuzzy, does it encourage the reader to fill out my contact form or call my phone number, would my audience feel compelled to share this on their Facebook, would my audience tell their friend about this article because it was so good)

A lot of SEO's have fallen prey to a set list of things they think they should be doing on a monthly billing cycle. They created a package for their client or boss so they could justify what they do. As a result, the client or boss expects certain tasks to be done in exchange for the monthly fee or salary the SEO is getting.

The problem with this method is that SEO is not just an art or a science – it's a blend of both. One cannot simply say that they need the following every month:

1. 1 Press Release
2. 10 Blogs
3. 25 Link Requests
4. 4 YouTube Videos
5. 2 New Webpages
6. 1 Email Newsletter
7. 1 Infographic

That would make it merely a science or a math equation and almost anyone could figure it out and be doing his/her own SEO. Of course there are some analytical things to pay very close attention to. And there is the competitive analysis that needs to be done and assigning weight to certain links and so forth. But SEO is also an art.

There is a certain amount of psychology that goes in to all of this. Understanding what our business or website does and how it applies to our target audience. Furthermore, how we can get them to engage with us and send us money. There really is no longer a simple A + B = C.

On Page Content

To be somewhat systematic, I would suggest that you begin with your core content. This would be your Home page and your top level Category pages. Think about which pages you have in your main navigation of your site. These should be your first focus.

Go through each page, one by one. Actually read what you have on your website. Funny as it may sound, many people have never even done this.

How does it read? How does it make you feel when you are reading it? Do you find yourself wanting more? Do you feel like this website has the most authority on this subject? Do you feel bored when you are trying to read the content? Could it broken down in to smaller pieces by using bullet points or sub headings?

Now go read two of your biggest competitors pages on the same topic. The ones that are ranking number one and number two for those keywords you want to have. How does their content read? How does their content make you feel?

You are probably starting to see a pattern here. Good! When

creating content, you need to put yourself in the same position you did when you got in to business. You told yourself that you were going to start a business that does what you do because you felt like you could do it better than anyone else in your industry. Right? That's exactly the mentality you have to have when creating content. If you believe you are the best at what you do, then show your audience in your content.

What makes your company unique or different? What do you do that no one else does in your industry? How is your product, customer service or formula better than everyone else? This is what you need to be displaying within your content. And this is exactly what the search engines want to see.

You could spend countless hours writing long documents about things that everyone already knows or they can find on Wikipedia. Or you could focus on what you do differently or better than the other guys. There is no need to get confused or overwhelmed here.

After you go through and corrected your core content and made it dazzle, it's time to move on to your inner pages and do the same thing. Along the way you may find certain pages that are not that necessary. Excellent! Time to do something with them. Maybe even get rid of them.

Keep in mind that deleting a page is not necessarily a bad thing.

The search engines look at your sitemap and determine what your site is about based on how all the pages content fits together. So if you have a very thin page that isn't really that relative to what you do, it's probably time to consolidate that content on to another page or get rid of it all together. Just be sure not to forget to do a 301 redirect.

You would be surprised how many pages you could consolidate on a large website to create new pages that actually start to make some sense. You can also save that old content and find a way to repurpose it later somewhere else like a blog post.

The main idea behind auditing your site's main content is to paint a clear picture for the user. You want them to have an excellent experience and know exactly who you are, what you do, what products and services you have, where you are located and how to contact you. Don't ever forget to think just like the big guys. They don't put up pages of content just to put up pages. Every single piece of content they produce has a purpose.

Once you have completed the entire main portion of your website's content, it's time to move on to auxiliary pieces, like your blog.

You will want to use a similar protocol when going through your blog posts. I would recommend setting up a spreadsheet and listing all the blog posts in your asset inventory.

From there I would try to put them in to categories. Keep your eyes open for blog posts that are very similar and could be consolidated.

Next you will want to see if you have any analytical data on these pages and posts. You certainly don't want to disrupt and existing traffic that you are getting by moving or deleting them.

A very good idea for old worn out blog posts is to repurpose them. Especially if they were getting traffic at an earlier time, you will want to try to revive them somehow. Possibly try rewriting them or adding some fresh images to them. Also be sure to get links to those posts and push them out on your social media.

If you have a blog on your site that is very old, chances are that you have a lot of spammy content still hanging around. You will want to identify these and get rid of them as soon as possible. Once again, don't forget to do 301 redirects.

A final note about on-site content: Spam sucks – get rid of it. If you don't, Google will do it for you. There is absolutely no reason to have bad content on your site anymore. There are no more tricks or tactics that work for spoofing the search engine. If you can think it up, chances are someone else has already thought of it too. And Google likely has already implemented something to find it, flag it and penalize it.

Off Site Content

Now let's talk about off-site content. As I mentioned earlier, there are a tremendous amount of places to publish content. Many of them are very good and healthy SEO. The bottom line is that you need to always keep your target audience in mind. Let's go over a few of them here.

YouTube is an excellent resource for producing content. I am a big advocate of this site for so many reasons.

The biggest problem I see with people using YouTube is they simply do not produce content. You have to actually go out with a camera and film something before you can put something up on this site! Sounds trivial, but it's the facts. So the only way to put this site in to your content strategy is to take your phone or your camera or whatever and go get some video footage.

Then you will want to do some sort of editing to make it look proper for your audience. Remember who the people are who you want to see these videos.

You will want to make sure that you are not trying to produce some feature film length video. No one wants to watch anything for more than 40 seconds. People are simply just too busy these days. And they really don't care about all the mumbo-jumbo you think you need in the video. Keep it simple and you have a better chance of being rewarded.

The same rules apply for video content. You will need to make sure that it is either education, informative or entertaining. Look around at some of the videos out there these days. You can easily tell which ones are effective and which ones are not by simply looking at the amount of views.

Remember that videos are a lot like images. The search engine can't really "see" them, so you need to effectively use the proper title, description and tags to guide the engine.

There are plenty of books and websites about YouTube and optimizing for it. I strongly recommend you start using this site if you aren't already.

Blogs are an excellent way to produce new content without interrupting the main area of your website. I have seen a lot of SEO's out there that talk about having to blog a certain amount of times per week, per month, etc. So what is the right answer? Simple. If you have something that is worthy of being written, published and read – then it's the perfect time to blog. Otherwise, save your time.

Once again, the last thing you want to do is start producing more web spam in the universe. More so, why would you want that spam attached to your optimized site? You take a big risk of the search engine lowering your overall authority and trust ranking by putting this junk on their radar.

Press releases are an excellent tool for getting real news out to the public. But let's make sure we are following the guidelines here. Just like all the other forms of content, there is no reason to publish something just to publish something. If it's not really news or newsworthy, then you are better off not putting it out there.

You may have read somewhere that press releases are good SEO. Let me clarify that for you: Good press releases are good SEO.

Once your company or your website truly has something that is newsworthy, it is time to produce a press release. Until then, it's probably best to save your time and money.

Infographics are something that came on scene a few years back. They seem to be very sexy and cool. Personally I like them, but I have one big problem with them: the user experience.

Have you seen most of these infographics? They are ridiculously long and offer an insane amount of data that could never be consumed by a human in one setting.

The best form of infographics I can ever remember were the ones that the USA Today newspaper has used for decades. Their graphics were short, sweet and to the point. I was able to wrap my head around what they were trying to show me. I consumed their content.

I would encourage you to use more infographics if they are fitting for your audience. Be sure to keep in mind that you are not trying to create the infographic of all infographics. You are trying to produce a piece of content that will quite possibly engage your audience to take a call to action and send you money for something.

eBooks are a great way to offer valuable information to your audience. The neat part about them is that you don't have to find a publisher or spend a lot of money to put one together. Actually it can be as simple as writing it in Microsoft Word and saving it as a PDF.

These can be excellent tools for several reasons. First you can use them to encourage visitors to fill out a contact form in order to receive the free download of your eBook.

Next you could utilize them to repurpose some old content.

Maybe you have a bunch of old pages that don't quite fit the new look of your website structure. Don't throw out that old content! Clean it up a little, put it together somehow, do whatever you need to do – and publish your first eBook!

There really isn't any right or wrong way to do an eBook. I believe there are several reasons people produce them and many more reasons why people read them. Lots of people like them because they are generally short and simple. They can download them quickly on to a device and read them while in a waiting room or on a flight.

I would certainly encourage you to explore all types of off site content. The possibilities are absolutely endless. Think about where your audience is and try to carve out a niche to reach them where they are. Try to do a quick study on what you are looking to produce for. Maybe you will need to buy a small book, read another website or watch a video on How To Write a Newsletter or How To Start Your Own Podcast. Either way, the main idea is to come up with an idea and simply begin to produce the content.

Always be testing different things. See what works and keep doing it. If it doesn't work, don't fret. Simply try to repurpose that content on a different medium. You would be surprised how certain pieces of content that seemed worthless can be converted to a different platform and become a hit.

Local SEO and Content

In closing about different types of content, I want to be sure to talk about Local SEO. Many people might find it interesting that I put this part here, but I can assure you that your optimization for Local has a lot to do with content.

One of the first things you must do for your Local SEO is to make sure you accurately depict your business information on your own website. This should be done in a fashion that makes it easy for the visitor as well as the search engine to find.

The most obvious and common way this is done is to have a Contact page on your website. It may seem redundant, but I must suggest that you ensure all the contact information on your website is correct and up to date. It might surprise you how many websites have incorrect information on them.

Another common format used for this information is to put it in the footer or sidebar area of your site. Maybe you are a business that relies solely on customers that come to your office or shop. You might want to consider having this information a little more front and center. Just keep in mind who your visitors are and where they would expect to see this information.

Next you will want to think about all the things your visitor might want from your contact information. Do they need directions? Do you have multiple locations and they need a way to find the one closest to them? Do you have a service or delivery area? Might they want to know what hours you are open? Or how they can quickly schedule an appointment? All these things matter to your target audience, therefore they should matter to you. Make their experience with your company as seamless as possible.

After you have cleaned up the Local SEO on your site, you will want to move on to the other websites that hold your company's information. Sites like Google My Business, Yelp, Yellow Page sites, City Search, etc. You will want to ensure that each of these sites has accurate information about your business or website and that it matches the data on your site.

Another reason I feel this fits in to the Content portion of this book is because each of these listings typically has areas to do little write-ups about your business. Many of them also have review portions of their site. Both of these are considered content and you will want to effectively manage it.

As far as the About section on these sites, it's a good idea to make sure you have a couple of well-written write-ups ready. After you claim these listing, you will want to set up all the phone number and address etc. Be sure to add nice images including your logo and pictures of your shop or facility where possible. Use nice and flowing keywords in your descriptions and be sure to write a good Call to Action if you can.

Encourage your customers to leave reviews on these sites. And be sure to respond to all reviews as soon as you can. People like to be acknowledged. Reach out to the people who left you great reviews and thank them. Maybe offer them a coupon for their next visit. Empathize with someone who might have left you a less than desirable review. See if there is a way you can make it right in hopes of them editing the review.

Reviews are unbelievable content that are used by many. Think about purchases or dining decisions you have made based on them. Reviews have made their way in to the hands of the search engines and are shown front and center for many businesses. Some have even speculated that they have a strong weight on the results page of local business for Google and others.

Overall the main idea is to be very mindful with your content. Don't be limited to just the content on your website. Think outside the box and explore where your audience might be. Always be testing and trying new things. Don't be afraid to spot a failure and try to turn it around in to a success.

For quite a while in the SEO world, we used to hire outside writers to produce a lot of our content. This, of course worked for many years until the Panda penalty. I don't want to take anything away from the really good quality writers out there. There are many of them available and you should consider using them where necessary. But I will say a few words about those "other" writers. Quite honestly – they are just not that good. You shouldn't expect anyone to know enough about your topic than yourself or your staff.

So that brings me to my next suggestion about writing content. Either do it yourself or get creative and find ways to include staff members, C-Level executives, brand ambassadors or even your customers. Who could be better at writing about your business? Don't trust your content to some unknown author who you hired to simply fill a gap. If it's worth writing, it's worth writing well.

The only thing you can truly do wrong with content is: not think about the user. If you keep the visitor top of mind, you will win.

Be a brand. Be an authority. Be the leader in your industry. If you are producing content that engages your audience, they will share it. Quality content has always mattered.

CHAPTER 5: LINK ATTRACTING

Often times you will hear people refer to this as Link Building. I believe that is an old term. Quite frankly I like to call it Link Buying, which is a major foul in search engine optimization these days. For years many people got away with "building" a ton of links to their sites and they were rewarded for them. As with anything else in the SEO world, any form of manipulation of the search engine rankings will always be caught and penalized.

If you get nothing else from this chapter: Don't Buy Links!

Penguin

Now to the address the next Google penalty I discussed earlier in the book: Penguin. From now on, you will hear it all the time about how you need to get links. And how you need to get good links. People will try to sell you the latest way to get links to your site. Who knows, you may even fall for it.

But ultimately what is the best or proper way to get links?

Quite simply, I am not going to blow a bunch of pixie dust around and share some crazy insider secret about where all the good links are and how to get them. It just isn't there. You can search high and low and read all the Black Hat SEO websites you want – you wont find them. You might run in to quite a few scams or con artists, but you certainly will not find a way to game the search engine with some secret, extra saucy links that no one knows about.

Are links still good SEO? Absolutely! They are one of the still remaining factors that you can use to send signals to the search engine. You just don't want to be thought of as trying to manipulate your ranking with the use of links. Conversely you want to make sure that you have good links pointing to your site that are relative and helpful for your audience.

Chances are you already have several links to your website. It's a good idea to take an inventory of the existing links to your site. This can be done by using one or more of the following tools:

1. majestic.com
2. ahrefs.com
3. Google Webmaster Tools
4. Moz.com
5. Smallseotools.com

Once you see your overall link profile, you will want to take inventory of it to sort out the good from the bad. I recommend using an Excel spreadsheet and try using some of the factors addressed by some of the above-mentioned tools. Things like Citation and Trust should be considered. The higher the number, the better.

Once you have your list separated in to the "good" links and the "bad" links, you can now start to find a way to start removing the bad ones. This can be a never-ending process for some sites as they may have been around for several years and other SEO's or agencies may have had their hand in previous link building.

You should use your spreadsheet to keep track of all the websites you reach out to. Try to tactfully and professionally request them to remove the link. You can find examples of emails that have been effective to do this.

Be sure to make notes about each link and if they respond to you. Some will ask for a "ransom" to remove your link. Don't pay them, just make a note. Some may never respond or you won't find any way to contact them. Not to worry, just keep writing the details of each one.

Once you have reached out to all of your "bad" link websites and have either gotten a response or not, gather up all your notes in to one text document.

Disavow Or Remove Links

In an effort to not have these links be held against you, you will want to utilize the Google Disavow Tool to try and get them removed from your websites equation. This process is not guaranteed to work, but it is at least worth the effort and it has worked for many websites. Be very careful with this tool. And be sure to read up a little bit before setting sail in to the unknown. It is a tool that is simple enough for anyone to use, but you will want to make sure you are doing it right so you don't end up with any major mistakes.

If you have a website that has entirely too many links to inventory and control, you may want to consider using a link removal service. Just be sure to use a reputable one and know that it may get expensive.

Some experts have debated on whether or not to "start over" with a new website if a link profile is too bad. That is only a decision you can make. Personally I believe almost any website can be recovered with enough effort and passion. But the choice is yours if you feel your site may have went too far over the years.

So What Now?

OK now let's assume that you have cleaned up your link profile and you only have mostly "good" links. Just how do we go about getting more "good" links? Getting other websites to link to you is about earning them. That means you will have to work for them.

In the past, people used to fire off a bunch of emails to other webmasters and say things to entice them to link to their site. Those became very annoying of course and generally go unnoticed now. Then there are other sites that have a link submission form on them where you can easily type in your information and be included on their site with a link to yours. And there are plenty of services out there that will gladly charge you a fee and get thousands of sites to link to yours. All of the above are bad ideas. You will not get any good links from doing these things.

The only way to get a proper good link is to produce content that is linkable. You may want to write that last sentence down. If you are building content with authority, people will want to link to you.

Think in terms of how social media websites work. While you scroll through your news feed on Facebook, you are given the option to Like, Comment or Share on any given post. Chances are you scroll for quite some time until something grabs your attention. You may stop and read it or look at it, and ultimately you might click Like. You may even find an article or a video that is so funny or interesting that you decide to click Share. Imagine if you clicked Like on everything in your news feed. Facebook and all your friends might start to think that you are nuts. And pretty soon your Likes, Shares and Comments become rather meaningless.

That's exactly how "good" links work. People like to link or share content that is good. In turn this sends the right signal to the search engine that there are other people on the web that find your content worthy of a link.

Links really are about content. If your content isn't great, chances are your links won't be either. But if your content is awesome, there will be a tremendous amount of good links coming your way. It's just like anything else these days. Celebrities and rock stars don't get popular and make a lot of money because people don't like them. They get there because people love them and are always talking about them to their friends and recommending their songs and movies. This is exactly how the web works!

In addition to great content, it is a good idea to reach out to what we call Friendly Sites. Chances are you have already identified several in your industry. These are usually niche directories or blogs that are well written. Contact them and see if there is any way you can be of help to them. Quite possibly you might be able to provide them with some new fresh content. In exchange they might offer to link back to your site.

Maybe there is a way you could become an ongoing writer for an industry magazine. And through that be able to put a link in your byline.

You will want to make sure you are continually building relationships within your space and finding new, creative ways to produce links for your business.

Try getting involved with Help a Reporter Out (HARO). You can sign up to be available as an authority in your industry. And when reporters or journalists are seeking interviews for a story they are doing, they will contact you. You may end up gaining a killer link from a big news station!

Depending on your industry, you may want to get more involved with sites like AngiesList.com or Yelp.com. There are many sites where you might find great success by being a trusted authority in your industry and people will find it wise to link to your site.

Don't forget about all the links that you have in your Local SEO listings. Make sure they are all current and correct.

As a general rule, directory submissions and social bookmarks are bad. I would recommend that you simply stay away from them. I would also advise against Guest Blogging. Be sure to not get involved in any purchasing of links, link exchanges, or link schemes. These will get your site instantly shot down.

A Final Note on Links:

PageRank is basically gone, or going away. Google announced this a few years back. So don't concern yourself with trying to get high PR sites anymore. Just look for relative sites and start to cultivate a relationship with them.

Focus only on trying to achieve natural links that real people from real websites give to you. Don't try to manipulate or game the system. It just won't work.

Spend more time making sure your content is the best. Share your great content with others and offer to provide content for sites that might be of help to your website.

CHAPTER 6: SOCIAL MEDIA

As of late, social media has become the rave. Everyone wants to "go viral" with something. It is often times perceived as a cheap or free way to reach multitudes of people in a short amount of time.

Some have even gone so far as to say that social media is the new SEO. I disagree with that statement completely. Social media has absolutely nothing to do with the on page factors of your website, up to and including your links and your content. As of the writing of this book, there is not foreseeable way that social media will replace SEO.

Social media is another marketing channel that should be used. It is an opportunity to engage with your target audience as well as your existing customer base and brand advocates. It's a great way to promote certain goods and services to increase revenues, if done properly.

Depending on your industry and your target audience, social media can be a big hit or you might miss the mark completely. If you are trying to sell mobility chairs or hearing aids, you might have to think outside the box a little more than a company trying to market video games or the latest movie release.

Just because you might think your industry doesn't need to be on social media, don't throw it out all together. You may be missing out on a few major benefits to your company.

Let's say you are in the business of selling heavy equipment. What if all your customers are only municipalities and state governments? Why would you need to have a Facebook page? Simple. You can utilize your Facebook page as another way for people to find and contact you. Certainly there must be other state and local government agencies that might need your equipment. What if one of your customers from 20 years ago forgot how to reach you and now they need to buy from you again?

Another reason you would want to set up your Facebook page is to encourage reviews. These reviews might foster future sales by being the one thing that helped someone make a decision.

How about being able to show the world that you are doing great things for your community? You could use this platform to post updated events you are contributing to. Or even write and publish helpful articles for issues that are relatable to the general public. Someone might even pick up that article and send a link back to your website.

So regardless of the business that you are in, you should definitely make sure you are using social media. Each case will certainly be different, but you only stand to benefit from having it.

From the beginning you will want to make sure that you set up all your social profiles. I would advise you to set up and claim them on all the social sites you can find. You won't necessarily want to use all of them, but it is important that you have them. You might find a use for an obscure one down the road. And it would not be good to find out that your desired name is already taken.

When setting up your profiles, try to protect your brand and your name(s). Try to match up your profile URL's amongst all the social sites, if possible. This will help with the user experience in the long run.

It can be quite confusing to your target audience if your URL's look something like this:

- twitter.com/abcflowers

- facebook.com/abcflowerco
- pinterest.com/abcflowerschicago
- instagram.com/abc_flowers

You and your audience will find it much smoother to set them up like this:

- twitter.com/abcflowers
- facebook.com/abcflowers
- pinterest.com/abcflowers
- instagram.com/abcflowers

Imagine being able to print your business cards or window wraps and only having to say: Follow Us: /abcflowers

Social media builds brands. Depending on your industry, you may likely never sell a single item directly from a social media site. As of recent, some of the social media sites are implementing a Buy Now or other Call to Action buttons in to their sites. I can see where this might have a big impact on certain businesses, especially those who are heavily dependent on social sites. But for a lot of businesses the focus will need to remain on building brands, managing reputation and reviews and interacting with customers and advocates.

Although it has been debated enough already, social media can have an effect on SEO. The debates probably stem from those who are seeing direct impacts versus those who are not.

When a company posts something on a social media site and it gets a tremendous amount of engagement, it simply cannot be ignored by the search engines. They are very aware of trends and heavy web activity to certain things. And the attribution of the content that was posted will certainly be given to the originator of the content.

Moreover there are additional signals that will occur as a result of something going viral. Let's say you post up a funny, original video about Taylor Swift riding a Harley Davidson. All of the sudden people start to like, comment and share your video. Chances are there will be a buzz around the search engines of people looking for the "Taylor Swift Harley Davidson" video. It is likely that your site will come up in the search results (or your YouTube channel, if that's where you posted it.) Search engines pay attention to the entities that are producing great content. They reward those content producers.

Social media can and should be used to push out your content to the masses. But that shouldn't be the only thing you are doing on social sites. The last thing you want to do is bore your audience with always posting about yourself. Remember that people want to be informed, educated or entertained. Nothing more and nothing less.

You are trying to build and manage your reputation here. Do you want people to think that you only try to sell things to them? They are going to get pretty tired of that pretty quick and stop following you. You need to be clever. Be yourself in so many ways. No one likes a brand talking AT them. They want to see some form of personality and have a brand talking WITH them.

You want to build up a listener base for your brand. Make people want to see what you are going to post next. Have them on the edge of their seats.

Maybe you are challenged with finding more people to follow you. Spread the word to your customers about following you on Facebook and Twitter. Let them know that you occasionally post coupons and specials exclusively on those sites. Share helpful information with them that will help them in their daily lives.

If you are thinking about your audience and how you can help them, then they will be more receptive to you when you post something about your business.

In order to get people to engage with you on social media, you have to create content that is engaging. Put yourself in their shoes every now and then. If you were them, would you share that post or video with your network of friends and family?

Usually the answer is no.

But if you create a really helpful video about how to do something around their house, they might just share it. Or if you post a really great coupon or offer, you might get a great response from your audience.

Social media is an excellent tool if done properly. Alternatively it can be a complete bust or waste of time when you only think about yourself and your business. You have to always be thinking about your audience and how you can be of service to them through these sites. No one wants to hear how great your company is or how wonderful your products are all the time.

People want something useful that they can share with others. People like to discover things first. They like to be the originator amongst their friends. They share things because they think it will make them look cool. What does that mean for you? You have to give them something that they can share that will make them look cool!

Tips for Social Media:

Here are some of my tips for being successful on social media:

1. Don't be annoying
2. Ask a lot of questions
3. Have a purpose
4. Don't over share
5. Quotes are getting old
6. Images and Videos are in right now
7. Be a leader
8. Be the authority in your space
9. Don't do what everyone else is doing. Stand out in the crowd
10. Leave people wanting to see what you are going to post next
11. Pay attention to what people are saying
12. Communicate with your audience. Don't let them do all the talking
13. Pay attention to different time zones. Know when your audience will read your posts.
14. Always grow your networks
15. Never miss a chance to help someone

CHAPTER 7: PAY PER CLICK AND PAID ADVERTISING

When it comes to SEO, it's time to put up or shut up. You have to be walking to walk and producing the best sites with the best content. Only the strong will survive. And the only alternative to good SEO is paid marketing (PPC), which can get severely expensive.

If done properly, SEO becomes an asset to the business, not an expense. Ongoing salaries or agency fees for proper SEO can eventually be attributed to the right lines in a profit/loss statement and a ROI can be calculated. Paid online marketing will continually rise over time and will become a hindrance to the overall profit/loss of a business.

Whereas it may be preposterous to try to create a hypothetical example, I will do so for the sake of learning. In this example I will create two companies that need online marketing: ABC and XYZ.

Let's say that ABC decides to use PPC as their only form of online marketing. They determine to spend exactly $4000 per month in their budget. And for the sake of assumption, let's say ABC gets a tenfold return on their investment every month for the first year. This would yield them $480,000 revenues on $48,000 marketing investment.

The next year some of ABC competitors see what's going on and decide to jump in on the action. They begin to compete with ABC for their PPC ads, thereby driving up the cost per click (CPC) on each of their ads. ABC now has to decide whether to adjust their CPC or to adjust their overall budget in order to maintain the level of revenues they have been generating.

If ABC decides to adjust their CPC, they will ultimately get fewer clicks. As a result, they will get fewer conversions and revenues on that same marketing spend of $4000 per month.

If ABC decides to increase the overall marketing spend to compete, then they obviously find themselves with a higher cost per acquisition too. They ultimately find a rising cost in order to continue experiencing the same revenues.

Chances are ABC will no longer be experiencing tenfold return on investment (ROI) after the first year.

Now let's look at XYZ, a company that decides to do all their online marketing via organic SEO. They establish a similar budget of $4000 per month for their efforts. For the sake of argument, we can assume that the first few months of SEO yielded $0 return.

Then around the third or fourth month XYZ starts to experience a small ROI. Likely they are breaking even in their current months. Over the next few months, XYZ begins to see their ROI increase by a couple thousand dollars month over month. By month 8 or 9, they are seeing returns in the neighborhood of $20-28,000 on their $4000 monthly marketing spend.

The trends continues as their good SEO efforts begin to really pay off around month 10 and on in to month 12. All together they are able to look back at the end of the first year and see a return of $241,000 in revenues on their $48,000 investment.

This seems like a far cry from the $480,000 that was produced by ABC in their first year, right? Sure it does – until the second year comes along.

XYZ begins year two with continuing to only spend exactly $4000 per month on their SEO efforts. However the trend continues to grow upwards on the returns. Assuming they hit their stride and peak performance of $60,000 per month in returns.

That would yield a total of $720,000 revenue on XYZ's $48,000 marketing investment. And chances are that ABC had to continue to throw more money at the PPC market in order to stay competitive. Yet they still were not able to hit $720,000 in revenues unless they spent significantly more on their marketing budget.

All that begs the question: Should you avoid PPC marketing all together? To put it in to just one word - NO. PPC has its benefits in certain circumstances.

I believe that a company should be utilizing PPC in their marketing efforts, so long as they are calculated.

Any business needs revenues to sustain and keep moving forward. SEO takes time in the beginning. Therefore I believe that a company should prepare to spend an allocated budget in the beginning to get started.

Unfortunately Google doesn't simply recognize that you launched a new site and it has proper SEO and decides to rank you on the first page for all your desired keywords. That would be like opening a new pizza shop that is the best pizza ever and having its location hidden from the general public, expecting everyone to flock to your business.

Just like in the real world, you have to do some marketing spend and get the word out to begin getting your opening traffic and revenues.

Another example of where I feel PPC is useful is to protect your brand. If you new company is called Chicago Pizza World and someone types that in the search box only to return Pizza Hut, Gino's East and Little Caesars – you probably want to protect your business buy allocating some of your marketing budget to that keyword.

Once again, look at what the big guys are doing. Apple, Amazon, Sears – you name it. They wisely protect their brands using PPC. This can avoid losing potential customers who are looking specifically for you and might get lost in a competitor's site that is ranking better than you for that keyword.

PPC has gotten even more competitive over the past few years. This is largely because SEO got tough. A lot of so-called SEO's jumped ship and started selling PPC management.

They are the ones who are always looking for the quick and easy buck. I am certainly not suggesting that everyone who provides PPC management are not good at online marketing. There are many reputable individuals and companies out there. If you are seeking someone to manage your PPC, I would caution you to be very careful with your selection process.

Keep in mind that hiring someone to manage your PPC means that you have to give them somewhere around 10% of your marketing spend. So if your PPC budget is $100,000 per month, plan on writing another check to them for $10,000 each month. And if you plan on doing your own PPC, you should probably be reading a different book and learning really fast how to manage your campaigns.

Many people that prefer PPC would consider it the opposite of SEO. Actually it should coincide with SEO most of the times, for many reasons. Both sides of online marketing should be working together and sharing intelligence with one another.

And the idea of SEO is to create a long-term asset that will reduce the PPC budget for the company. PPC is on and gone. Once a click happens, your money is spent. When SEO is done correctly, you retain an asset that will continue to work for you.

PPC is very super, ultra-competitive. It is an auction format where the highest bidder wins. So if a competitor wants your spot, they simply flash more cash. And if you want to keep up, you have to dig in to your war chest for more money.

CHAPTER 8: SCHEMA AND STRUCTURED DATA

What is structured data? In the past we used meta data as a way to tell the search engines what a page was about. We would try to give the engine a basic idea of a page by stuffing in meta tags (keywords) and meta descriptions (little summaries). Then the search engines started to get a bit wiser. This because the SEO's of the time were simply trying to manipulate the system by putting in all the high-powered keywords within their meta data in order to rank for such terms.

What structured data is designed to do is show the search engine certain pieces of content within your website that are potentially important to users. With this code, we are able to somewhat help the search engine provide specific answers to specific queries.

In current day, you have probably seen structured data in action and you might not have even known it. You can usually find one of the best examples of structured data is going to Google and searching for one of the latest movies that is in theaters. Before you do this exercise, think about what you would expect to see in the results. Now go ahead and search for the title of this week's blockbuster film.

Depending on your location and device, you will likely see the following for that film:

- Show times
- Theaters playing the film
- A way to watch the trailer easily
- Reviews of the film
- Short description of the film
- A list of the main actors and directors
- Release date
- And much more

Additionally by clicking on one of the show times, you are probably able to purchase a ticket relatively easy.

Think about what you would want to see if you searched for a movie. All the above data is exactly what you were thinking, right? Google successfully provided the answer to your query by understanding your intent.

How did they do this? By pulling structured data from websites that are providing this data. If you click on several of the links or you search for a couple other recent films in theaters, you will likely see a pattern of the websites Google is pulling from. Usually sites like Fandango, IMDB, Rotten Tomatoes, Wikipedia and so on. All of these sites are currently providing structured data for these movies.

Let's try another example. Go to Google and type in the word: Pizza. Notice the results that are displayed for you. You may see some pizza shops or franchises listed with their reviews, average costs, and maybe even some descriptions of their locations and offerings.

Then you will likely see some information being pulled from Wikipedia describing what pizza is. And most recently Google has begun pulling data from the USDA to provide you with Nutrition Facts for the food you query.

What does this mean for you? That depends on what your website is about, including the industry you are in. You might not think that this structured data has anything to do with you or your business. Think again.

More and more search results are being pulled from this microdata. Keep in mind that Google wants to be the source of all answers to all questions. Some of the Godfathers of SEO have even suggested that the "ten blue links" on Google pages might be going away soon. That would mean that we are no longer looking for traffic to our websites from Google, but rather a possibility of conversions directly from the search results pages.

So even if you are trying to optimize a website for a local mechanic's shop, you still have a lot of opportunity to utilize structured data.

What can be structured? Almost anything can be put in to this format. At a minimum you will want to tell the search engine the basics of your business or entity.

For example, you will want to make sure that your Name, Address and Phone Number (NAP) are coded in schema. This way the search engine can use this information to produce it within their search results where necessary. Think about a person who is driving around town and thinks about your services. They hold up their smartphone and ask Google to show them a mechanic. In return, Google shows them some of the closest mechanics to their current location. They will even show them reviews, address and likely a phone number. Most importantly, Google might even provide them with directions to your business if you have added the proper schema to your website.

Schema can be used for so many different entities around the web. You should make sure to add schema to the important areas of your website that would be of the most service to your target audience.

Here are some other ideas of items you should consider adding schema for:

- Videos
- Key Staff Members
- Different Locations
- Hours of Operation
- Events
- Product Listings
- Recipes
- Restaurant Menus
- Reviews from Customers
- Different Creative Works
- Show Times
- Flight Times
- Reservations or Tickets
- University Affiliations
- Type of Business or Service
- Specific Location Details
- Specific Product Details

The opportunities are endless for certain websites. For others it can be as simple as adding in a few extra lines of code.

You should familiarize yourself with the website schema.org and explore which entities you should be providing structured data for. Always be thinking about your target audience. Go out of your way to really put yourself in their shoes. Think about what they are doing and what they are thinking when they search for your product, service or business. Be creative and think of a way that you may be able to reach them directly through the search results. You might just find a way to easily beat your competition and earn a first page ranking. Moreover you may even begin to gain instant conversions directly from the search page without ever getting any real traffic to your site!

There really can't be enough said about structured data. A lot of the foresight is leaning towards it. Much of the new and upcoming technology is using it to steer users in ways we can only imagine.

Now is the time to set yourself up for future success and be a part of what was only considered science fiction in the recent past. Imagine how this data is being used and processed with today's technology. How is a voice search processed? How does Google use your previous search behavior to provide you with the exact answer to your query?

The overall essence of structured data is for us to tell machines how to produce specific answers to human beings based on the intent of their query. To provide people with the least barrier to entry of being converted in to our customer.

Using the movie example from earlier, this means you will be able to ask your phone to search for a movie title, read or watch a brief on it, see a couple of reviews quickly, and be able to buy tickets instantly. You will probably never have to even navigate a single link or website. With a simple acknowledgement by voice or fingerprint, you can complete the purchase and be on your way to the theater! This whole process started because of structured data.

For the beginner or novice coder, this can be a daunting process. It is not the easiest thing to do-it-yourself. Even if you have to find someone and pay them a few extra dollars, it could be well worth it for your website or business in the long run.

Structured data could very well be your trump card versus your competition.

CHAPTER 9: TAKEAWAYS AND SUGGESTIONS

The underlying theme of this book is all about the user experience. For far too long, search engine optimization was thought to be about manipulating a machine (the search engine.) The web as we know it has rapidly evolved since it's inception. Most people have evolved with it and they have a higher expectation of what they can get from the common search.

Whether they think about it consciously or not, when a person performs a search – they have a specific intent or motivation behind that query. Our job as search marketers is to meet that intent or motivation with the best possible answer. We have to provide the ultimate content that is completely unique to anywhere in the world.

It is simply not good enough to take some information from another website or Wikipedia entry and spin it in a different way. The search engines have become much smarter and faster. They consider a page written like that is merely a citation (or a vote) for that original page. Have you been wondering why you are seeing more Wikipedia entries in the search results the past couple of years? It's because all that spam content that was drafted from a spun version of their content is proof to Google that Wikipedia is the authority. Therefore they deserve the first page result for your keyword.

Often times that is the biggest challenge for many industries: to produce content that is more authoritative and unique than a simple Wikipedia entry.

Today, SEO demands that you are a thought leader. You can no longer merely take what someone else is doing and try to repeat the process. You have to build your website completely clean. You have to review all your code and ensure your markup and tags are done properly. You have to find what makes you or your business unique and portray that to the world in a different way.

There are no more tricks or tactics that will work, at least for very long. The search engines have spent literally billions to combat this nonsense. They are smarter than they have ever been and will only continue to become wiser in the coming years. Our job is to cooperate with them and produce high quality content on clean, easy to use websites.

There cannot be enough said about the user experience. Our websites have to be available to our target audience whenever, wherever, and however they want to access them. If we want to earn their business, we have to show them we are the absolute best in our business. We can only do that by producing high quality, unique and authoritative content. Whether that be text, images, videos or even our posts on social media sites. SEO's have to consider every aspect of the buying process.

Search marketers need to find ways to engage users. Make people a part of your content and conversations. There is no such thing as making a site that says: "Hey! We're the best! So fill out our contact form and we will call you."

The Internet has become second nature to most of the human race. They have all become savvy to the old techniques. People know they can expect the best answer from a Google search. As a result, this is why Google awards the top spots to the best possible answers.

Sure, it all sounds quite simple when you read it. The truth is that it can be a lot of hard work. This is usually why so many people decide to step away from SEO and either hire someone or jump on the paid advertising wagon.

Chasing the algorithms is hard work. Looking for the latest trick or tactic and trying to implement it is even more work.

If you take the time to do it right from the beginning, you will never have to take corrective measures. Or if you have to make all those corrections, it only has to be done one time.

Moving forward from here and choosing to do things the right way, you will never have to worry about the latest updates or penalties.

When building new pages or websites, or even making changes to existing ones, you should always consider doing so in a natural fashion. The same goes for content on and off site.

Think about how it would look if someone (or the search engine) shows up tomorrow and everything is completely different than it was yesterday. Out of nowhere, you magically have a new website and all new content and like wildfire you are suddenly producing multiple YouTube videos and gaining a huge following on social media sites.

Chances are that person (or the search engine) will be in a little bit of shock. It just won't seem natural to them. It could cause them (or the search engine) to put up their little red flags and not trust you as much.

So by taking the time to do everything systematically and organically, you will show people (and the engines) that you are a real person behind a real website that is doing real SEO. As a result, you will earn more trust and set yourself up for better ranking positions and more conversions.

I am sure you have heard or read it a thousand times by now, but SEO takes time. It is not an overnight fix. There are no gurus or agencies out there that can wave a wand and make all your Google issues go away in the blink of an eye. Even the best of the best of SEO's will have to do all these same techniques to your website and content.

Don't be fooled by anyone who says the can get you on the first page or provide you with some amazing content that no one else can.

Hopefully you have chosen to give this a shot at doing most of this yourself. You are in for an awesome journey. Search engine optimization is both an art and a science. And now more than ever it is about psychology.

You have complete control of your destiny online, but you have to follow the rules. Many of these rules (or best practices) have not changed since day one. The moment you take a break or make a wrong decision to find a faster way, your website will likely face consequences. Alternatively, if you stay the course and focus on the user experience, you will gain big wins for the long term.

To come to a conclusion of this book, I want to take a moment to remind you that this book is not the "be all, end all." I strongly encourage you to read other books, watch videos, try new things on your website (or a test website). You should go to a training course or a conference of some sort in the near future. Reach out to others in the SEO space online through social media or online communities.

All of the knowledge in this book is nothing new or ground breaking. It is a compilation of the following:

- Training from 7 years of Internet marketing conferences
- Relationships built within the SEO community over the years
- Starting websites from nothing and making an earnest living from them for almost 9 years, even selling 1 online company for a huge gain
- Performing numerous SEO trainings for companies nationwide
- Continuing to practice the art of SEO with my personal clients ongoing

You too can become a good sound SEO practitioner. Be patient and always learning.

I wish you continued success and encourage you to make the web a better place for the user.

CHAPTER 10: TOOLS OF THE TRADE

I could give you examples of so many different tools that are out there for just about anything and everything. To be honest with you, I believe this list is quite exhaustive. Nonetheless I believe that everyone that is involved with search engine optimization should be utilizing certain tools that enable you to:

- Increase Performance
- Save Time
- Save Money

Additionally many tools help with tracking and testing to learn from our experiences along the way and become better at what we do.

When exploring this list, keep in mind that this is not everything that is out there currently, but it's a really overwhelming good start for you. Additionally I am sure there will be more tools created in the near future that you can get your hands on.

I am in no way connected or being paid by any of the following tools or websites. I do not advocate for any particular tools. You will find which ones you are most comfortable with and which ones provide you with the most value for what you are trying to accomplish.

One more thing to keep in mind is that many of these tools do more than one thing or the website listed offers a multitude of tools (or a suite). But for simplicity's sake, I have categorized many of them.

Core Basics

Google Webmaster Tools

www.google.com/webmasters/tools

Bing Webmaster Tools

www.bing.com/toolbox/webmaster

Google Analytics

www.google.com/analytics

Open Site Explorer

www.moz.com/researchtools/ose/

Robots Generators

www.tools.seobook.com/robots-txt/generator/

www.mcanerin.com/EN/search-engine/robots-txt.asp

internetmarketingninjas.com/seo-tools/robots-txt-generator/

Sitemap Generators

www.xml-sitemaps.com

www.web-site-map.com

www.freesitemapgenerator.com

www.xmlsitemapgenerator.org

Schema Creator

www.schema-creator.org

Structured Data Testing Tool

www.google.com/webmasters/tools/richsnippets

Competition/PPC/Keywords:

Wordtracker

www.wordtracker.com

Wordle

www.wordle.net

Wordstream

www.wordstream.com/keywords

Spyfu

www.spyfu.com

Majestic SEO

www.majestic.com

SEM Rush

www.semrush.com

Keyword/Topic Suggestion Tool

www.sg.serpstat.com

Similar Web

www.similarweb.com

Authority Labs

www.authoritylabs.com

ScrapeBox

www.scrapebox.com

Ubersuggest

www.ubersuggest.org

Link Checkers and Monitoring:

Ahrefs

www.ahrefs.com

Majestic SEO

www.majestic.com

Screaming Frog

www.screamingfrog.co.uk/seo-spider/

Link Removal:

Removeem

www.removeem.com

Content Calendars:

Brandeo

www.brandeo.com/2012_Marketing_Calendar_Template_Free_D
ownload

DivvyHQ

www.divvyhq.com

Gather Content

www.gathercontent.com

Publish This

www.publishthis.com

Content DJ

www.contentdj.com

Super Awesome Content Strategy Worksheet

www.axzm.com/super-awesome-content-strategy-worksheet

Local Tools:

Yext

www.yext.com

Local Site Submit

www.localsitesubmit.com

Best of The Web

www.botw.org

Local.com

www.local.com

Local SEO Checklist

www.localseochecklist.org

Getlisted.org (now Moz Local)

www.moz.com/local

Free Review Monitoring

www.freereviewmonitoring.com

Social Media Tools:

Hootsuite

www.hootsuite.com

Buffer

www.bufferapp.com

Tweetdeck

www.about.twitter.com/products/tweetdeck

Social Bro

www.socialbro.com

Bit.ly URL Shortener

www.bit.ly

Trackur

www.trackur.com

Page Speed Tools:

Page Speed Insights

www.developers.google.com/speed/pagespeed/insights/

Pingdom Speed Test

www.tools.pingdom.com

GT Metrix

www.gtmetrix.com

Max CDN

(content delivery network)

www.maxcdn.com

Cloud Flare

www.cloudflare.com

Amazon AWS CloudFront

www.aws.amazon.com/cloudfront

CDN.net

www.cdn.net

Full Suites of Tools:

Moz Tools

www.moz.com/tools

Internet Marketing Ninjas

www.internetmarketingninjas.com/tools/

Hubspot

www.hubspot.com

Raven

www.raventools.com

SEO Book

www.tools.seobook.com

SEO Site Checkup

www.seositecheckup.com

Woorank

www.woorank.com

SEO Tool Set

www.seotoolset.com

Submit Express Tools

www.submitexpress.com/tools.html

Small SEO Tools

www.smallseotools.com

WebCEO

www.webceo.com

Other Helpful Tools:

Word2cleanhtml

(convert Word Documents to HTML)

www.word2cleanhtml

Jpeg Mini

(reduce image sizes)

www.jpegmini.com

Copyscape

(duplicate or stolen content check)

www.copyscape.com

SEOQuake Toolbar

www.seoquake.com

Yoast

www.yoast.com

Crazy Egg

www.crazyegg.com

Firebug Web Development Tool

www.getfirebug.com

Help a Reporter Out

www.helpareporter.com

Wayback Machine

www.archive.org/web

TubeMogul

www.tubemogul.com

Slideshare

www.slideshare.net

Alexa

www.alexa.com

Compete

www.compete.com

Outsourcing Services:

Fiverr

www.fiverr.com

eLance

www.elance.com

oDesk

www.odesk.com

Freelancer

www.freelancer.com

Content Writing Services:

Text Broker

www.textbroker.com

Scripted

www.scripted.com

CopyPress

www.copypress.com

Crowd Content

www.crowdcontent.com

Ongoing Community and Learning:

Search Engine Journal

www.searchenginejournal.com

Search Engine Land

www.searchengineland.com

Search Engine Watch

www.searchenginewatch.com

Thread Watch

www.threadwatch.org

SEO Round Table

www.seroundtable.com

Webmaster World

www.webmasterworld.com

SEO Chat

www.seochat.com

Moz

www.moz.com/community

Search Marketing Conferences

SMX

www.searchmarketingexpo.com

SES Conference

www.sesconference.com

MozCon

www.moz.com/mozcon

Content Marketing World

www.contentmarketingworld.com

PubCon

www.pubcon.com

BlogHer

www.blogher.com/conferences

NMX / Blogworld

www.blogworld.com

Conversion Conference

www.conversionconference.com

South by Southwest

www.sxsw.com

Philip Cory

ABOUT THE AUTHOR

Phil Cory has successfully been in the internet industry for the past 10 years. He has started several successful online ventures and even recently sold his largest website to one of the largest conglomerates in his industry. Phil regularly trains SEO and does consulting throughout the United States.

www.ingramcontent.com/pod-product-compliance
Lightning Source LLC
Chambersburg PA
CBHW071221050326
40689CB00011B/2399